BAOFENG RADIO BIBLE 2025

A Comprehensive Handbook for Easy Mastery and Reliable Communication When It Matters Most

ISAAC J. STONE

Copyright Notice

No pert of this book may be reproduced, written, electronic, recorded, or photocopied without written permission from the publisher or author. The exception would be in the case of brief quotations embodied in critical articles or reviews and pages where permission is specifically granted by the publisher or author. Although every precaution has been taken to verify the accuracy of the information contained herein; the author and publisher assume no responsibility for any error or omissions. No liability is assumed for damages that may result from the use of the information contained within.

All Rights Reserved Isaac J. Stones ©2025

INTRODUCTION

CHAPTER ONE

BAOFENG RADIOS
WHAT IS A BAOFENG RADIO?
BASIC RADIO CONCEPTS
LEGAL CONSIDERATIONS

CHAPTER TWO

GETTING STARTED WITH YOUR BAOFENG RADIO
CHOOSING THE RIGHT MODEL
CHOOSING THE RIGHT MODEL FOR YOUR NEEDS
ESSENTIAL ACCESSORIES
UNBOXING AND SETUP

CHAPTER THREE

MASTERING THE BASICS
PROGRAMMING YOUR RADIO MANUALLY
GETTING FAMILIAR WITH THE KEYPAD AND DISPLAY
STEP-BY-STEP GUIDE TO MANUAL PROGRAMMING
USING CHIRP SOFTWARE FOR PROGRAMMING
BASIC OPERATION

CHAPTER FOUR

FREQUENCY AND CHANNEL MANAGEMENT
UNDERSTANDING FREQUENCY RANGES
VHF VS. UHF: KEY DIFFERENCES
CHOOSING THE RIGHT FREQUENCY FOR DIFFERENT SCENARIOS
PRACTICAL TIPS FOR USING VHF AND UHF
EXAMPLES OF COMMON FREQUENCIES AND THEIR USES
SETTING UP YOUR RADIO FOR OPTIMAL FREQUENCY USE
TROUBLESHOOTING FREQUENCY ISSUES
SCANNING AND MONITORING
WHY SCANNING AND MONITORING MATTER
ADVANCED SCANNING OPTIONS
PRACTICAL SCANNING TIPS

Troubleshooting Scanning Issues
Saving and Organizing Channels

CHAPTER FIVE

Advanced Settings and Techniques
CTCSS/DCS for Privacy
Repeater Access and Offset Frequencies
Dual-Watch and Dual Receive Functions

CHAPTER SIX

Communication Protocols and Best Practices
Radio Etiquette and Protocols: Speaking Like a Pro
Understanding Signal Reports and Signal Strength
Emergency Communication
Tips for Optimizing Communication Clarity and Reliability

CHAPTER SEVEN

Troubleshooting and Maintenance
Common Issues and Fixes
Regular Maintenance
Software and Firmware Updates
Troubleshooting Flowchart

CHAPTER EIGHT

Practical Applications and Real-World Scenarios
Outdoor Adventures: Hiking, Camping, and Wilderness Scenarios
Urban and Group Use: Staying Connected in Urban Environments and Event Coordination

CHAPTER NINE

Expanding Your Skills and Knowledge
Connecting with the Radio Community
Next-Level Licenses and Equipment
Ongoing Learning Resources

CONCLUSION

INTRODUCTION

Imagine you're on a remote hiking trail with no cell service, or perhaps you're in the middle of a storm, and traditional communication lines are down. How do you stay connected to your group, call for help, or get crucial updates? In moments like these, having a reliable means of communication can make all the difference, and that's where the Baofeng radio steps in. Often overlooked as an amateur tool, Baofeng radios are, in fact, powerful communication devices that can be indispensable in emergencies, survival scenarios, and everyday situations alike.

The Baofeng Radio Bible was written with one primary goal: to empower you to use your Baofeng radio confidently and effectively, especially when reliable communication is critical. Mastering your radio doesn't just mean understanding buttons and settings; it means knowing when and how to use it in real-life situations. Whether it's an emergency that cuts off traditional communications or a hike where cell service is unreliable, this book will guide you through the essentials and the nuances of using your Baofeng radio, so you're prepared for any situation.

Who Will Benefit From This Book

This guide is tailored to meet the needs of a wide range of readers:

- **Beginners** who may feel overwhelmed by technical jargon and just want to get their radio up and running,
- **Hobbyists** curious about amateur radio communication as a fun and useful skill,
- **Preppers and survivalists** who understand that when systems fail, being self-sufficient in communication is vital,
- **Outdoor adventurers** who value dependable communication as a safety tool on their journeys,
- And anyone interested in **learning to stay connected** regardless of where they are or what circumstances arise.

Baofeng radios have become beloved tools among amateur radio enthusiasts and everyday users worldwide. Their appeal lies in their powerful functionality and affordability. From the well-known UV-5R to more specialized models, Baofeng offers a variety of options for communication across frequencies, making it possible to broadcast, receive, and stay informed across considerable distances. Whether for personal preparedness, outdoor adventures, or a new hobby, these radios have proven that reliable communication can be accessible and affordable.

In the following chapters, you'll find a structured guide that covers everything from the basics of operating your Baofeng radio to advanced techniques for programming, maintaining, and troubleshooting it. By the end, you'll not only have the technical know-how but also the practical insight needed to use your radio confidently, ensuring you're prepared to communicate in any situation. Let's dive in and unlock the full potential of your Baofeng radio.

CHAPTER ONE

Baofeng Radios

In the world of two-way communication, Baofeng radios stand out for their power, versatility, and affordability. For newcomers, these handheld devices might look a bit intimidating, with multiple buttons, dials, and antennas. But don't worry! You are going to have a solid understanding of what a Baofeng radio is, key radio concepts you'll need to know, and essential regulations to ensure you're operating within the law.

What Is a Baofeng Radio?

Baofeng radios are handheld transceivers that allow you to transmit and receive signals over various frequencies. The Baofeng brand is widely known for its UV-5R model, which became a popular choice among amateur radio users, preppers, outdoor enthusiasts, and even professionals who need reliable communication tools. Let's explore a few popular Baofeng models to get a sense of their range and capabilities.

1. **Baofeng UV-5R**: The UV-5R is perhaps Baofeng's best-known model and a great starting point for beginners. It covers both the VHF (136-174 MHz) and UHF (400-520 MHz) frequency bands, which means it can communicate over both long and short distances. This model has a 4-5 watt output power, and while that might not sound like much, it can reach several miles in optimal conditions, especially if paired with a high-quality antenna.

2. **Baofeng UV-82**: The UV-82 is like the big sibling of the UV-5R. It has the same dual-band functionality but is often preferred for its enhanced durability and improved audio quality. With an upgraded speaker and microphone, it provides clearer sound and also features a slightly more powerful output at up to 8 watts. This model is ideal for situations where stronger signal transmission is necessary, like in hilly or forested terrain.

3. **Baofeng BF-F8HP**: The BF-F8HP is a high-power version of the UV-5R, boasting up to 8 watts of output power, making it one of Baofeng's most powerful handheld radios. This added power can extend communication range and signal strength, which is particularly useful in areas where other models may struggle.

Each of these models supports both **Simplex** (direct radio-to-radio communication) and **Duplex** (communication through repeaters) modes. These radios also allow users to program specific frequencies, which we'll cover in more detail later in the book.

Basic Radio Concepts

Understanding your Baofeng radio also means getting familiar with a few key radio concepts: **frequency, channels, bandwidth**, and **modulation types**.

1. **Frequency**
   Frequency, measured in Hertz (Hz), describes the number of oscillations a radio wave completes per second. Baofeng radios typically operate in the VHF (Very High Frequency) and UHF (Ultra High Frequency) bands. Here's a simple breakdown:
   - **VHF (136-174 MHz)**: These frequencies are ideal for open areas with few obstructions, such as flat countryside or plains. VHF signals are longer and can travel farther in clear conditions.
   - **UHF (400-520 MHz)**: UHF frequencies work well in built-up or forested areas because they can penetrate obstructions better, though they generally don't travel as far as VHF signals.

   Example: Imagine you're on a hiking trail in a heavily forested area with a friend. If you both set your Baofeng radios to a UHF frequency, you'll likely get a stronger signal than with VHF due to the UHF's ability to navigate through trees and rocks.

2. **Channels**
   Channels are essentially slots or preset frequencies that you can program into your Baofeng radio. This makes switching between commonly used frequencies easier, so you don't have to manually tune each time. Think of channels like a contact list on your phone, allowing you to quickly call up specific frequencies.

3. **Bandwidth**
   Bandwidth refers to the width of the frequency band being used, typically measured in kilohertz (kHz). In Baofeng radios, you may encounter both "narrow" (12.5 kHz) and

"wide" (25 kHz) settings. Choosing the right bandwidth can help improve clarity and avoid interference:
- **Narrow Bandwidth (12.5 kHz)**: Preferred in crowded areas, narrow bandwidth is good for maintaining clear signals in dense frequency environments.
- **Wide Bandwidth (25 kHz)**: Suitable for areas with less radio traffic, where signal clarity won't be impacted by other frequencies.

4. **Modulation Types**

Modulation is the process of encoding information on a radio wave, with **Frequency Modulation (FM)** being the standard for Baofeng radios. FM provides clear sound and is generally less prone to static compared to **Amplitude Modulation (AM)**, which is used more for commercial broadcasts (think AM radio stations). Baofeng radios use FM because it works well for voice communication, ensuring you can be clearly heard even in areas with some interference.

Legal Considerations

Operating a Baofeng radio can be a fun and practical hobby, but it's important to stay within the rules. Regulations vary by country, so it's essential to understand the specific guidelines in your area. In the United States, the **Federal Communications Commission (FCC)** sets these rules.

1. **Licensing Requirements**: To legally use certain frequencies on your Baofeng radio in the U.S., you need a license:
   - **No License Required**: If you're using your Baofeng radio on General Mobile Radio Service (GMRS) or Family Radio Service (FRS) channels, no license is typically needed.
   - **Amateur Radio License (Technician Class)**: For broader access to VHF and UHF frequencies, an FCC Technician Class License is required. This license allows you to access a wide range of frequencies and power levels up to 1500 watts. You'll need to pass a basic exam covering radio knowledge and etiquette.
2. **Regulations on Power Output and Frequency Usage**: Baofeng radios come with various power settings (like 1 watt, 4 watts, and 8 watts), and it's important to note that higher power settings aren't always legal to use without a license. The Technician Class

License, for example, allows for higher output, but unlicensed users must stick to approved FRS frequencies and lower power settings.

3. **Proper Radio Use**: Using a Baofeng radio also means respecting radio etiquette and other users. This includes keeping transmissions short, avoiding crowded frequencies, and respecting emergency communications. Here's a handy list of etiquette tips to keep in mind:
    - **Identify Yourself**: Use your call sign if licensed, or provide a clear identifier (e.g., "Hiker on Channel 5") if you're operating without a call sign.
    - **Keep It Brief**: Long messages clog up the airwaves. Stick to concise, clear communication.
    - **Listen Before Speaking**: Ensure the channel is clear before beginning your transmission.
    - **Emergency Use**: Channels like 146.520 MHz are dedicated to emergencies. Only use them for urgent matters unless specified otherwise.

Example: If you're camping in a remote area and your friend is a few miles away, you could use an open frequency within the UHF range. But if you hear other users on the channel, you might move to another frequency to avoid interference, allowing everyone to communicate smoothly.

Now, you have a foundational understanding of what a Baofeng radio is; the basic concepts behind how it works, and the legal guidelines to keep in mind.

CHAPTER TWO

Getting Started with Your Baofeng Radio

Welcome to the start of your hands-on experience with Baofeng radios! Whether you've chosen a model to carry on a hiking trip, prepare for emergencies, or just experiment with new communication tools, this will guide you through selecting the best model, identifying essential accessories, and setting up your device step-by-step. By the end, you'll be ready to turn on your Baofeng radio with confidence and know exactly what each button and feature is for.

Choosing the Right Model

1. Baofeng UV-5R: The Classic and Affordable All-Rounder

Overview:

The UV-5R is Baofeng's most popular model. It's compact, affordable, and has a range of features that make it versatile for various uses. Its dual-band capability (VHF and UHF) allows you to access different frequencies, and it's suitable for general communication needs, both for beginners and more advanced users.

Key Features:

- **Frequency Range:** VHF (136-174 MHz) and UHF (400-520 MHz)
- **Output Power:** 4-5 watts
- **Battery Life:** 1800 mAh battery for extended usage
- **Programmable:** Can be programmed manually or with CHIRP software

Ideal Uses:

- **Outdoor Enthusiasts:** The UV-5R's range and power are sufficient for short-range communication during hiking, camping, or other outdoor adventures.

- **Urban and Family Use:** Its affordability makes it easy to equip multiple family members, making it ideal for group communication in urban environments, such as events or crowded areas.
- **Beginners:** The simple interface and manual programming options make this a good starting point for those new to ham radios.

Considerations:

The UV-5R doesn't have many advanced features, and while it's a solid choice for beginners, more experienced users may find it limited for specialized applications like long-distance or heavy-duty use.

2. Baofeng UV-82: Enhanced Durability and Ergonomics

Overview:

The UV-82 builds on the UV-5R but is known for its improved ergonomics and dual push-to-talk (PTT) button, which allows you to switch between two channels seamlessly. It also has a sturdier build, making it more suitable for outdoor and rugged use.

Key Features:

- **Frequency Range:** Same as UV-5R (VHF and UHF)
- **Output Power:** Slightly higher, around 5-7 watts
- **Dual Push-to-Talk (PTT):** Allows for quick switching between two frequencies
- **Enhanced Durability:** Designed to withstand harsher environments better than the UV-5R

Ideal Uses:

- **Preppers and Emergency Situations:** The UV-82's durable build makes it more reliable in emergency or rough conditions.
- **Group Events and Volunteer Coordination:** Dual PTT is handy for monitoring and switching between team channels without reprogramming.

- **Outdoor Activities:** For activities like hunting, fishing, or camping, the UV-82's build and dual-channel operation provide more flexibility than the UV-5R.

Considerations:

The dual PTT feature can take some getting used to, but it's valuable if you frequently switch between channels. However, for those focused on budget, the UV-82 is slightly more expensive than the UV-5R.

3. Baofeng BF-F8HP: High Power and Extended Range

Overview:

The BF-F8HP is the "high power" version of the UV-5R, boasting an 8-watt output. This higher wattage extends range and signal clarity, making it ideal for users who need more reliable coverage over longer distances. It's a popular choice for those who want a step up in power without moving to a significantly more advanced model.

Key Features:

- **Frequency Range:** VHF and UHF
- **Output Power:** 8 watts for extended range and better signal quality
- **Improved Battery:** Comes with a 2000 mAh battery for longer usage times
- **Display and Interface:** Upgraded screen and more intuitive menu layout

Ideal Uses:

- **Outdoor and Rural Use:** The extra power is beneficial in rural areas or mountainous regions where obstacles may interfere with signal range.
- **Emergency Communication:** Its enhanced power can be critical during emergencies where clarity and range are essential.
- **Advanced Users or Preppers:** For those looking to improve signal quality and reach without upgrading to professional radios, the BF-F8HP provides an accessible option.

Considerations:

The BF-F8HP is bulkier and slightly heavier than the UV-5R and UV-82 due to its larger battery and higher power output. Users should also be cautious with battery usage at 8 watts, as it can drain faster when transmitting continuously.

4. Baofeng GT-3WP: Waterproof and Weather-Resistant

Overview:
The GT-3WP is one of Baofeng's more rugged models, offering waterproof and weather-resistant capabilities. It's IP67 rated, meaning it can withstand temporary submersion and is ideal for wet or harsh environments.

Key Features:

- **Frequency Range:** Standard VHF and UHF
- **Output Power:** 5 watts
- **Weatherproofing:** IP67 rating for dust and water resistance
- **Robust Build:** Reinforced casing for durability

Ideal Uses:

- **Water-Based Activities:** If you're near water or planning activities like fishing or boating, the GT-3WP's waterproof feature protects it from splashes or even accidental drops in water.
- **Harsh Environments:** This model is suited for areas prone to rain, dust, or extreme temperatures, such as deserts or humid climates.
- **Long Outdoor Stays:** The reinforced casing and weather resistance make it reliable for prolonged outdoor use.

Considerations:
The GT-3WP tends to be more expensive and heavier than non-waterproof models. Also, its battery life is comparable to standard Baofengs, so bringing an extra battery might be wise for extended use in remote areas.

Choosing the Right Model for Your Needs

To help you make a clear choice, here's a quick comparison table:

Model	Power	Special Features	Ideal Use
UV-5R	4-5W	Basic, affordable	Beginners, urban use, family comms
UV-82	5-7W	Dual PTT, rugged build	Group events, outdoor, emergency prep
BF-F8HP	8W	High power, better range	Rural/remote, advanced users, emergency
GT-3WP	5W	Waterproof, IP67 rated	Water activities, extreme environments

Example Scenarios:

1. **Hiking with Family**: The UV-5R is a great choice here if you need multiple radios for a family group. However, if you're going into mountainous terrain, the extra power of the BF-F8HP might be beneficial.
2. **Organizing a Volunteer Event**: For coordinating teams across a large event area, the UV-82 is ideal due to its dual PTT, allowing easy switching between communication channels.
3. **Preparing for an Emergency**: Preppers might find the UV-82 or BF-F8HP best for emergencies where durability or enhanced power is needed. For flood-prone areas, the GT-3WP could add a layer of reliability.
4. **Boating or Fishing**: The GT-3WP is designed for wet environments and can withstand water exposure, making it perfect for water activities.

Essential Accessories

To make the most of your Baofeng radio, consider adding a few key accessories. These can enhance your radio's performance, extend its battery life, and protect it from damage.

1. **Antennas**

   The stock antennas on Baofeng radios work fine for short distances, but for greater range and clarity, upgrading to a longer or more specialized antenna can make a big difference.
   - **Nagoya NA-771 Antenna**: This is a popular option among Baofeng users. It's longer than the stock antenna, improving range and reception without being too bulky.
   - **Magnetic Car Antenna**: Perfect for mobile use, this attaches to your car's roof, allowing for stronger transmission and reception while driving.

   Example: If you're planning to use your Baofeng radio for road trips or across wide, open areas, a magnetic car antenna will give you a clearer signal than a standard handheld antenna.

2. **Batteries**

   Most Baofeng radios come with a rechargeable battery, but having extras on hand is a smart idea, especially for longer outings.
   - **Extended Battery Pack**: Some batteries offer twice the capacity of the standard pack, lasting up to 24 hours on a single charge.
   - **AA Battery Adapter**: For situations where charging isn't an option, this adapter lets you use regular AA batteries to power your radio, a great backup in emergency scenarios.

   Example: If you're heading into a remote area without reliable access to power, bringing an extra battery pack or AA adapter can keep you connected much longer.

3. **Chargers**

   Baofeng radios usually come with a standard desktop charger, but there are other options for on-the-go charging.

- o **Car Charger**: Plugs into your vehicle's power outlet so you can recharge while driving.
- o **USB Charger Cable**: Allows you to charge your radio through any USB outlet, making it easy to top up your battery with a power bank.

4. **Protective Cases**

Investing in a durable case can protect your radio from scratches, drops, and other damage.
- o **Soft Shell Case**: Lightweight and flexible, soft cases are a good option for casual use.
- o **Hard Case**: Offers maximum protection and is ideal for rough environments or packing in a backpack.

Example: If you're using your Baofeng radio outdoors in rugged conditions, a hard case can prevent accidental damage, ensuring your radio stays operational.

Unboxing and Setup

Getting your Baofeng radio out of the box and ready for use is exciting. Here's a detailed setup guide to ensure you know your device inside and out.

1. **Unboxing**

    When you first open your Baofeng radio package, you'll likely find:
    - o The radio unit
    - o Detachable antenna
    - o Battery pack
    - o Desktop charger
    - o Belt clip
    - o User manual

    Tip: Carefully read through the manual to familiarize yourself with your specific model's unique features.

2. **Assembling Your Radio**

   Follow these steps to set up your Baofeng radio for the first time:
   - **Attach the Antenna**: Screw the antenna onto the top of the radio. Make sure it's securely fastened to avoid loose connections.
   - **Insert the Battery**: Slide the battery into the back of the radio until you hear a click. This indicates that it's properly seated.
   - **Attach the Belt Clip**: If you plan to carry the radio on your belt or bag, use a small screwdriver to fasten the belt clip to the back of the radio.

3. **Charging the Battery**

   Before using your Baofeng radio for the first time, it's a good idea to fully charge the battery:
   - Place the radio in the desktop charger.
   - Connect the charger to a power source, and allow it to charge until the indicator light turns green, which means it's fully charged.

4. **Familiarizing Yourself with Buttons and Functions**

   The buttons on Baofeng radios might seem overwhelming at first, but each serves an essential function. Here's a basic rundown, using the UV-5R as an example:

   Diagram 1: Key buttons and functions on a Baofeng UV-5R radio

   - **Power/Volume Knob**: Located at the top, this turns the radio on and off and adjusts the volume.
   - **PTT (Push-to-Talk) Button**: On the left side, this button is what you'll press to transmit your voice.
   - **A/B Button**: This lets you toggle between two pre-set frequencies or channels.
   - **Frequency Mode/Channel Mode Button (VFO/MR)**: Located below the screen, this switches between entering a specific frequency (Frequency Mode) and selecting a saved channel (Channel Mode).
   - **Keypad**: The numbered buttons allow you to manually enter frequencies, access the menu, and adjust settings.
   - **Flashlight**: Some models have a built-in flashlight, activated by a button on the side. It's a helpful feature in low-light situations.

- **Menu Button**: The Menu button lets you access the radio's settings, where you can adjust various parameters such as power level, squelch, and frequency offsets.

5. **Initial** **Settings**

Once you're familiar with the buttons, it's time to make a few basic adjustments:
- **Switch to Frequency Mode**: Press the VFO/MR button to enter Frequency Mode. This will allow you to manually input frequencies to get a feel for how the radio operates.
- **Adjust Squelch**: Use the Menu button to enter the squelch settings. Setting it to a medium level will reduce background noise without blocking weaker signals.
- **Set Volume and Test the Audio**: Turn the volume knob and press the PTT button to make a test transmission, ensuring you can hear yourself or a friend clearly.

Example: Say you're practicing with a friend. Have them tune to the same frequency, then press the PTT button to speak. If they're close by, you'll hear their voice come through clearly, indicating that both radios are set up correctly.

CHAPTER THREE

Mastering the Basics

Once your Baofeng radio is set up, it's time to move on to learning the core skills that will transform this tool from an intimidating device into a reliable companion. Let's cover essential skills, from manually programming your radio to using CHIRP software for quick setup. And also learn how to operate your radio effectively, from powering it on to using shortcuts that simplify your experience. Mastering these basics will prepare you for dependable communication when it counts most.

Programming Your Radio Manually

Why Manual Programming Matters

Programming your Baofeng manually lets you:

1. **Set Up Quickly in the Field:** Whether you're on a hike, at a remote site, or in an emergency, you can configure your radio without needing a computer.
2. **Adjust to Real-Time Needs:** For example, if you discover a new frequency to monitor or need to change privacy settings (like CTCSS/DCS tones), manual programming enables immediate adjustments.
3. **Build Essential Skills:** Getting familiar with the radio's interface will help you feel in control, even if you later rely on software for more complex setups.

Getting Familiar with the Keypad and Display

Before diving into programming, here's a quick overview of the keys you'll commonly use:

- **VFO/MR Button**: Switches between VFO (Variable Frequency Oscillator) mode, where you can select a specific frequency, and MR (Memory Recall) mode, where you access saved channels.

- **A/B Button**: Switches between the top and bottom displays on your screen, which can hold two different frequencies or channels.
- **MENU Button**: Opens the programming menu, allowing you to set options.
- **Arrow Buttons**: Used to navigate through menu items.
- **EXIT Button**: Exits the menu or cancels actions.
- **Numeric Keypad**: Allows you to directly enter frequencies, channel numbers, and other settings.

Step-by-Step Guide to Manual Programming

1. Entering Frequency Mode (VFO Mode)

Start by putting the radio into VFO mode, which lets you tune into any frequency:

1. **Press the VFO/MR Button**: This switches the radio into VFO mode, allowing frequency entry instead of accessing pre-programmed channels.
2. **Select the Display Line**: Press the A/B button to choose whether you're working on the top or bottom line.

2. Entering a Frequency

Once in VFO mode:

1. Use the keypad to type the frequency directly (e.g., "146520" for 146.520 MHz).
2. Confirm that the display shows the correct frequency.

3. Setting the Transmission Power Level

The Baofeng UV-5R and similar models allow you to set power levels for each frequency:

1. **Press MENU** and use the arrow keys to find **Menu 2: TXP** (Transmission Power).
2. **Select High or Low Power** using the arrow keys. High power can extend range, while low power saves battery life.
3. **Press MENU to confirm** and then **EXIT** to save.

Example: In a city or urban area, lower power is often sufficient and reduces interference; in rural settings, you might prefer high power for extended reach.

4. Setting Up Privacy Tones (CTCSS/DCS)

If you want to minimize interference or talk privately within a group, you can add privacy tones:

1. **Press MENU** and navigate to **Menu 11: R-CTCS** (Receive CTCSS) or **Menu 13: T-CTCS** (Transmit CTCSS).
2. Select the desired CTCSS tone from a list (e.g., 100.0 Hz).
3. **Press MENU** to confirm, and then **EXIT** to save.
4. Repeat for DCS if needed by selecting **Menu 10: R-DCS** (Receive DCS) or **Menu 12: T-DCS** (Transmit DCS) and choosing the desired DCS code.

Example: If your group sets 100.0 Hz on T-CTCS, all radios with that tone will only hear transmissions from others using the same tone, reducing cross-channel interference.

5. Saving to a Channel

Once you've entered a frequency and any additional settings like power or tones, it's time to save everything to a channel for easy recall:

1. **Press MENU** and go to **Menu 27: MEM-CH**.
2. Use the numeric keys or arrow buttons to choose an empty channel (e.g., Channel 1, 2, etc.).
3. **Press MENU to save** the frequency and settings to this channel.
4. **Press EXIT** to return to normal operation.

Example: You might save local emergency frequencies to Channels 1-5 for quick access in a crisis.

6. Verifying and Accessing Saved Channels

To verify that everything saved correctly:

1. **Switch to MR Mode** by pressing **VFO/MR**.
2. Use the arrow keys or keypad to cycle through your saved channels.
3. Check that each saved channel shows the correct frequency and settings.

Advanced Settings You Can Program Manually

In addition to basic frequency and tone settings, Baofeng radios offer a range of advanced options that can be programmed without software:

- **Scanning Mode**: Set up the scanning function to monitor multiple channels.
    - **Menu 16: SC-REV** lets you control the scan resume method—whether it pauses on an active channel or resumes scanning.
- **Bandwidth Selection**: Choose narrow or wide bandwidth to fit your frequency's characteristics.
    - Go to **Menu 5: W/N** to toggle between Wide and Narrow.
- **Busy Channel Lockout**: Prevents transmitting on a frequency if it's already in use, reducing interference.
    - Access this via **Menu 7: BCL**.

These advanced options offer more control over how your radio operates, letting you adapt settings for different scenarios.

Example: Programming a Local Repeater Channel

Suppose you want to program a local repeater channel. Here's an example sequence:

1. **Enter VFO Mode** by pressing **VFO/MR**.
2. **Enter the Repeater Output Frequency** (the one you listen to) using the keypad (e.g., 146.850 MHz).
3. Set the **Offset Frequency** for repeater access:
    - **Press MENU**, find **Menu 26: OFFSET**, and enter the repeater offset (e.g., 0.600 MHz).
4. Set the **Offset Direction**:

- o Go to **Menu 25: SFT-D** and choose **+** or **-** depending on the repeater's setup.
5. If required, add a **CTCSS Tone** for repeater access:
    - o Use **Menu 13: T-CTCS** and select the correct tone (e.g., 100.0 Hz).
6. **Save the Repeater Frequency and Settings to a Channel** using **Menu 27: MEM-CH**.

With these steps, you'll have your Baofeng set up to access the repeater, allowing you to transmit and receive across greater distances.

Final Tips for Manual Programming Success

- **Practice Regularly**: Familiarity comes with repeated use, so practice programming channels even if you don't need them immediately.
- **Create a Cheat Sheet**: Write down important menu numbers and settings for quick reference.
- **Be Patient**: Manual programming can feel tedious, but with practice, it becomes second nature.

Manual programming transforms your Baofeng from a simple device to a tailored communication tool, capable of supporting you in any situation. The more you practice, the easier it becomes to adjust settings on the fly, an essential skill for effective, reliable communication when it matters most.

Using CHIRP Software for Programming

Manual programming can be tedious, especially if you have many channels to set. CHIRP software simplifies this process, letting you input frequencies and settings on your computer and upload them to your radio in one go. Here's how to get started with CHIRP.

Step-by-Step Guide to Using CHIRP

1. **Download and Install CHIRP**
    - o Head to the CHIRP website and download the software that matches your operating system (Windows, macOS, or Linux).
    - o Install CHIRP following the instructions on the website.

2. **Connect Your Radio**
   - Plug a programming cable into your computer and the other end into your radio's headphone/microphone jack.
   - Ensure your radio is turned on, set to Frequency Mode, and that it's on a "quiet" frequency with no active transmissions.
3. **Download the Radio's Data**
   - Open CHIRP, go to **Radio > Download From Radio**, and select your radio model from the dropdown menu. This step downloads your radio's current settings to CHIRP, allowing you to make edits.
   - Follow any prompts, including setting the serial port if prompted.
4. **Edit and Add Frequencies**
   - In CHIRP, you'll see a grid with columns for frequency, name, tone, and other settings.
   - Click on an empty row to add a new frequency. Enter the **frequency**, **name** (label for the channel), **tone settings** (if using CTCSS), and any other details.
   - You can easily add multiple channels, set custom names, and configure repeater offsets with ease.
5. **Upload Data to Radio**
   - Once you've added your desired frequencies, go to **Radio > Upload to Radio**. This transfers your edited settings to the radio.
   - Confirm the settings are correct, and then disconnect your radio once the upload completes.

Example Setup with CHIRP

Imagine you want to program a list of emergency frequencies and local repeaters. Here's how this would look in CHIRP:

- In CHIRP, create rows for each frequency, entering details such as frequency (e.g., 146.500 MHz for emergency channel), custom names (e.g., "Emerg"), and any CTCSS tones if necessary.

- Enter all frequencies in one session, then upload to your radio. In a few minutes, you'll have all emergency channels saved and labeled for easy access.

Basic Operation

Now that your radio is set up, let's walk through basic operation so you can start communicating confidently.

1. **Powering On and Adjusting Volume**
   - Rotate the **power/volume knob** at the top of the radio clockwise. You'll hear a click, and the radio will power on.
   - Turn the knob further to adjust the volume. It's best to set the volume at a mid-level initially and adjust as needed.

2. **Switching Between Frequency and Channel Modes**
   - Use the **VFO/MR** button to toggle between Frequency Mode and Channel Mode:
     - **Frequency Mode** lets you input frequencies directly and is useful for one-off conversations or testing frequencies.
     - **Channel Mode** allows you to select preset channels you've programmed, which is useful for regular or emergency contacts.

3. **Scanning Channels**
   - To initiate scanning, press and hold the **[*] Scan** button. This will cycle through available channels, pausing when it detects an active frequency.
   - Scanning can be helpful for monitoring activity in your area without having to manually switch channels.

4. **Using the A/B Button**
   - The **A/B** button toggles between the top (A) and bottom (B) display lines on your screen, allowing you to monitor two frequencies simultaneously.
   - This feature is especially useful if you want to monitor both an emergency frequency and a local repeater.

5. **Locking the Keypad**
   - To prevent accidental button presses, press and hold the **# Key** to lock the keypad. This is particularly useful when carrying the radio in a pocket or pack.

- To unlock, press and hold the **# Key** again.

Example Scenarios Using Key Shortcuts

Imagine you're hiking with friends, and you want to stay connected across two channels: a general chat frequency and an emergency frequency.

- **Use A/B Mode** to set the chat frequency on Channel A and the emergency frequency on Channel B.
- **Enable Dual Watch** mode so your radio will automatically monitor both frequencies and alert you if someone broadcasts on the emergency channel.
- **Lock the Keypad** after setting up to avoid accidentally changing frequencies while you move.

We have covered everything you need to start using your Baofeng radio confidently, from manual programming to CHIRP setup and basic operations. With these essentials under your belt, you're well-equipped to handle a variety of situations and connect with others.

CHAPTER FOUR

Frequency and Channel Management

With a solid grasp of basic operations, it's time to explore the world of frequencies and channels, a critical part of using your Baofeng radio effectively. Let me guide you through understanding VHF and UHF ranges, explain how to use the radio's scanning features, and show you how to organize channels to be prepared in any scenario. Mastering these areas will empower you to tune into the right communications when it matters, whether for casual use, emergency situations, or specific activities.

Understanding Frequency Ranges

In the simplest terms, a frequency range refers to the band of electromagnetic frequencies over which a radio can transmit or receive signals. Each range has different propagation characteristics, meaning it behaves differently depending on the environment, distance, obstacles, and weather conditions.

Your Baofeng radio typically covers two main ranges:

- **VHF (Very High Frequency)**: 136-174 MHz
- **UHF (Ultra High Frequency)**: 400-520 MHz

These ranges offer flexibility to adapt to various communication needs, from short-range indoor communication to longer-distance outdoor use.

VHF vs. UHF: Key Differences

Very High Frequency (VHF)

- **Frequency Range**: 136-174 MHz
- **Ideal For**: Outdoor use with minimal obstructions, long-distance communication in open areas.

- **Propagation Characteristics**: VHF waves travel farther over open terrain and penetrate vegetation, making them suitable for rural or outdoor environments.
- **Limitations**: VHF signals have difficulty penetrating buildings, metal, and dense urban environments with many obstacles.

Ultra High Frequency (UHF)

- **Frequency Range**: 400-520 MHz
- **Ideal For**: Indoor environments, urban areas with many obstacles, short-range communication with good clarity.
- **Propagation Characteristics**: UHF waves are better at penetrating walls, metal, and other solid objects, making them ideal for indoor and urban use.
- **Limitations**: UHF waves don't travel as far as VHF in open terrain and are more affected by environmental factors like foliage.

Choosing the Right Frequency for Different Scenarios

Choosing between VHF and UHF depends largely on the environment and your communication needs. Here are some practical scenarios to illustrate when each frequency range might be more effective:

1. Outdoor Adventures (VHF)

Imagine you're on a mountain hike with a group. You're likely in an open, rural area with few buildings or obstacles. In this case, VHF would be ideal because its lower frequency allows it to travel farther over open distances and through vegetation. You and your group members can stay connected across longer distances without signal loss, especially if you're spread out over hilly terrain.

2. Urban and Indoor Use (UHF)

Now, consider that you're attending a large event at a convention center with several floors and walls that can block signals. Here, UHF is the better choice. Its higher frequency enables it to penetrate walls and other obstacles, allowing you to communicate reliably even when you're not in direct line-of-sight. UHF's ability to handle interference from metal and concrete structures makes it perfect for city environments or indoor use.

3. Mixed Environments (Both VHF and UHF)

If you're in a mixed-use environment (e.g., in a small town where you'll be moving between open areas and buildings), you might find both VHF and UHF useful. Your Baofeng's dual-band capabilities mean you can switch between the two depending on your location. For outdoor parts of town, VHF may offer better range, while UHF could be more effective when you're inside buildings.

Practical Tips for Using VHF and UHF

Use High Ground for VHF

VHF waves travel best in open spaces and can benefit from elevation. When using VHF outdoors, try to communicate from higher ground if possible. This can extend your range and help you avoid obstacles that might block signals.

Optimize Your Antenna for UHF in Urban Areas

If you're using UHF in a city environment, a higher-gain antenna (such as a whip antenna) can improve clarity and range. UHF's shorter wavelength benefits from better-quality antennas, which can help reduce interference from surrounding buildings.

Consider the FRS/GMRS Frequencies on UHF

In the United States, UHF frequencies in the Family Radio Service (FRS) and General Mobile Radio Service (GMRS) are commonly used for short-range communications. Baofeng radios can be programmed to monitor these frequencies for local communication, especially in areas where others may be using walkie-talkies on these channels.

Examples of Common Frequencies and Their Uses

Here's a quick overview of some common frequencies within VHF and UHF bands that are often used in specific scenarios:

Frequency Range	Common Use Cases	Example
144-148 MHz (VHF)	Amateur Radio Bands (Ham), Emergency Comms	146.520 MHz
162.400-162.550 MHz (VHF)	NOAA Weather Broadcasts	162.550 MHz
400-470 MHz (UHF)	FRS/GMRS Channels, Public Service, Security	462.550 MHz
450-470 MHz (UHF)	Business and Industrial Communications	464.500 MHz

These examples highlight how specific frequencies are dedicated to certain uses. Monitoring the weather on VHF or staying connected on FRS/GMRS channels on UHF are straightforward and practical applications of frequency management on your Baofeng.

Setting Up Your Radio for Optimal Frequency Use

When setting up your Baofeng for either VHF or UHF, keep in mind the environment and intended range:

1. **Program Common Frequencies**: Program the channels you'll use most often (like emergency or weather channels) so they're ready to access when needed.
2. **Test Both VHF and UHF**: If you're unsure which frequency range is better, run tests in your environment. You might find VHF works best in the park, while UHF is superior indoors.
3. **Use Scanning and Dual-Watch Features**: Scanning both bands or enabling Dual-Watch (monitoring two frequencies at once) can help you stay connected across different ranges without constant manual switching.

Troubleshooting Frequency Issues

If you're experiencing issues with either VHF or UHF, try the following solutions:

- **Interference**: If you're getting interference on UHF in an urban area, consider moving to a higher location or switching to a different frequency.
- **Weak Signal on VHF**: If your VHF signal is weak, especially in mountainous or forested areas, try using a higher-quality antenna or communicating from elevated terrain.
- **Program Local Repeaters**: For both VHF and UHF, programming local repeaters can extend your range. Repeaters can be found in most areas and are often listed online by amateur radio groups.

Scanning and Monitoring

One of the powerful capabilities of Baofeng radios is their ability to scan and monitor multiple frequencies, allowing you to stay aware of activity on various channels without manually switching back and forth. Whether you're listening for emergency alerts, checking in on local weather channels, or keeping tabs on group communications, scanning and monitoring can keep you connected to the information you need.

Why Scanning and Monitoring Matter

Imagine you're hiking in a remote area with friends. You've programmed your Baofeng radio to scan emergency frequencies, local weather alerts, and a few channels for other hikers in the region. By enabling scanning, your radio will automatically cycle through these channels, stopping on any frequency with activity. This way, you don't need to constantly check each channel manually, you're alerted the moment any relevant broadcast comes through.

Scanning and monitoring are useful for:

1. **Emergency Situations**: Quickly catching alerts on weather, emergencies, or distress calls.
2. **Outdoor Adventures**: Staying connected to other group members or nearby radio users.

3. **Urban Monitoring**: Keeping tabs on family members, event attendees, or security channels.

Getting Started with Scanning on Your Baofeng Radio

Baofeng radios offer several scanning options, allowing you to select the type of monitoring that best suits your needs. Here's how to get started:

1. Enabling Basic Scanning

To activate scanning on your Baofeng, follow these steps:

1. **Switch to MR Mode**: Start by pressing the **VFO/MR button** to access Memory Recall mode, where you can scan saved channels rather than open frequencies.
2. **Press and Hold the *SCAN* Button (usually the *UP Arrow*)**: Your radio will begin scanning through all saved channels in memory.
3. The radio will pause on any channel where it detects a signal, allowing you to listen in.

To continue scanning after the channel becomes silent, simply wait a few seconds (if in *Time-Operated Mode*, TOS) or press the **UP Arrow** again to resume.

Advanced Scanning Options

Your Baofeng radio includes additional scanning functions to fine-tune what and how you monitor:

2. Adjusting the Scanning Resume Mode

You can select how the radio behaves when it encounters an active channel during scanning. There are three modes, set via **Menu 16: SC-REV**:

1. **TO (Time-Operated)**: The scan pauses on an active channel for a set time (usually five seconds), then resumes scanning.
2. **CO (Carrier-Operated)**: The scan pauses and remains on an active channel as long as it detects a signal. Scanning resumes only once the signal ends.

3. **SE (Search)**: Scanning stops when it detects a signal and remains on that channel until you manually restart scanning.

These options allow you to tailor the scanning experience based on your preference:

- **TO Mode** is great for casual monitoring of many channels, briefly pausing to catch transmissions but resuming quickly.
- **CO Mode** is ideal when you're actively listening and want to hear the full conversation.
- **SE Mode** is useful when you're waiting for a signal on a specific channel and don't want to miss it.

3. Scanning Specific Frequency Ranges

In addition to channel scanning, you can set up your Baofeng radio to scan within a specific frequency range. This can be useful for finding active channels in your area without pre-programmed frequencies:

1. **Switch to VFO Mode** by pressing **VFO/MR**.
2. **Enter a Starting Frequency** directly using the keypad.
3. **Activate Scanning** by pressing and holding the *UP Arrow*. The radio will scan through the frequencies in that range, pausing on any active ones.

Example: Let's say you're exploring a new area and aren't sure what frequencies are in use. You could start scanning in the 144-148 MHz VHF band (common for amateur radio) or 462-467 MHz UHF range (used by FRS/GMRS radios) to find local traffic.

Setting Up and Organizing Channels for Effective Scanning

For the most efficient scanning, organize your channels based on purpose and priority. This way, your Baofeng will monitor the frequencies that are most important to you:

1. **Group Channels by Purpose**: Set aside specific channels for different uses, such as emergency frequencies, family communication, weather alerts, and local repeaters.

2. **Limit Channels for Faster Scanning**: If you only need to monitor a few key channels, save time by only including those channels in memory.
3. **Prioritize High-Use Channels**: Place frequently used or high-priority channels in lower memory slots for faster access during emergencies.

Practical Scanning Tips

Here are some additional tips to make the most of your Baofeng's scanning feature:

- **Adjusting Squelch Settings (Menu 0)**: The squelch setting controls the radio's sensitivity to weak signals. Lower squelch settings allow you to detect weak or distant signals, though they may also pick up noise. Raising the squelch level can help eliminate background noise but may cause you to miss weaker signals. Test various squelch levels to find the right balance for your environment.
- **Use Dual Watch Mode**: Dual Watch (Menu 7) lets you monitor two channels simultaneously. This is especially useful if you want to keep an ear on a priority channel (like an emergency channel) while also scanning a secondary one.
- **Add CTCSS/DCS Tones for Privacy**: If you're scanning a channel used by multiple users (such as in a busy area or during an event), adding a CTCSS or DCS tone can help reduce interference from others using the same frequency.

Example Scenario: Using Scanning on a Group Hike

Imagine you're leading a group hike in a remote location. You've pre-programmed three essential channels:

1. **Emergency Frequency**: A channel for emergency communications (e.g., 146.520 MHz in the US).
2. **Weather Channel**: Local NOAA weather alerts, which automatically update in case of severe weather.
3. **Group Channel**: The primary channel for your group's communication.

To keep everyone safe and informed, set your Baofeng to scan these channels:

- **Start Scanning in MR Mode** to cover the emergency, weather, and group channels.
- Set the **Scanning Resume Mode to TO** so that the radio pauses briefly on any active channel but resumes scanning to monitor all three.
- Adjust your **squelch level to minimize background noise** while ensuring you can catch all relevant signals.

With this setup, you'll hear any emergency alerts, updates from your group, or weather changes in real time, all without having to manually switch between channels.

Troubleshooting Scanning Issues

If your Baofeng radio isn't scanning as expected, consider these troubleshooting tips:

- **Check Your Mode**: Ensure you're in MR Mode for channel scanning or VFO Mode for frequency scanning.
- **Verify Squelch Settings**: High squelch levels may prevent the radio from detecting weak signals, so adjust accordingly.
- **Channel Organization**: Make sure that your important channels are stored in the radio's memory and aren't skipped during scanning.

Saving and Organizing Channels

When you have multiple frequencies to manage, organizing them into channel lists will simplify usage, especially if you move between locations or handle different types of activities. This organization ensures that when it's time to switch frequencies, you can do so effortlessly.

How to Save Channels

Once you've manually programmed or used CHIRP to set a frequency, you can save it to a channel for quick access:

1. **Switch to Frequency Mode**
    - Press **VFO/MR** to go to Frequency Mode, where you can enter specific frequencies.

2. **Input and Save the Frequency**
   - Enter the frequency, press **Menu**, and select **MEM-CH** (Menu #27) to save it to a channel number.
   - Choose an open channel or overwrite an existing one by confirming your selection.

Organizing Channel Lists

Consider creating lists based on location, activity, or emergency scenarios. Here's how you might organize channels for different needs:

Example Lists

1. **Urban Use**:
   - **Channel 1**: Local repeater for city-wide communication.
   - **Channel 2**: Emergency services monitoring.
   - **Channel 3**: Group channel for urban exploration or events.
2. **Camping or Wilderness**:
   - **Channel 1**: Campground channel for group communication.
   - **Channel 2**: Emergency weather alert frequency.
   - **Channel 3**: Local ranger or park information frequency, if available.
3. **Emergency Preparedness**:
   - **Channel 1**: Local emergency broadcast.
   - **Channel 2**: Family or group check-in channel.
   - **Channel 3**: Neighbor or neighborhood watch channel for disaster coordination.

Using Labels and Names in CHIRP

When programming with CHIRP, you can label each frequency with a descriptive name, like "Emerg" for an emergency channel or "Camp" for your campsite frequency. This makes switching channels as simple as recognizing the label, saving time in critical moments.

Understanding frequency ranges, using scanning to monitor activity, and organizing channels, you're building a framework for successful, reliable communication. Whether you're camping, exploring a new city, or preparing for emergencies, you'll know exactly how to choose the best frequency, stay informed through scanning, and quickly switch to the right channels.

CHAPTER FIVE

Advanced Settings and Techniques

By this point, you've mastered the essentials of Baofeng radios, from powering on to scanning channels and organizing frequencies. Now it's time to dive into some advanced features that can elevate your experience, helping you communicate more efficiently and securely. Let's cover the privacy features like **CTCSS/DCS**, explore the powerful world of **repeaters and offset frequencies**, and unlock **dual-watch and dual receive functions** to monitor multiple channels at once. Mastering these settings and techniques will make you a versatile and confident communicator, ready for any situation.

CTCSS/DCS for Privacy

The Continuous Tone-Coded Squelch System (CTCSS) and Digital-Coded Squelch (DCS) are often described as "privacy" features, though they don't actually make your conversations private in the way encryption does. Instead, they help filter out unwanted transmissions and interference from others on the same frequency. Here's how they work and why they're useful.

Understanding CTCSS and DCS

- **CTCSS (Continuous Tone-Coded Squelch System)**: CTCSS sends a continuous sub-audible tone along with your transmission. Only radios programmed to listen for that specific tone will "open the squelch" and hear the transmission.
- **DCS (Digital-Coded Squelch)**: DCS works similarly to CTCSS but instead uses a digital code rather than an analog tone, offering more unique codes and potentially more interference resistance.

Imagine you're in a crowded area, like a busy park or an event, and everyone's using radios on the same frequency. By setting a CTCSS or DCS code on your Baofeng, you can effectively "tune out" other users who aren't using the same code. It's especially helpful in areas where many people use radios, as it keeps you from constantly hearing unwanted conversations.

Setting Up CTCSS and DCS on Your Baofeng

Here's a step-by-step example of how to set a CTCSS tone for communication with a group without interference from other radio users on the same frequency:

1. **Choose a Frequency**: Set your Baofeng to the frequency you'll use for communication.
2. **Access the CTCSS Menu**:
    - Press **Menu** to enter the settings.
    - Scroll to the **T-CTCS** (Transmit CTCSS, Menu #13) setting or **R-CTCS** (Receive CTCSS, Menu #11) if you only want to filter incoming signals.
3. **Select a Tone**:
    - Choose a CTCSS tone by scrolling through the available options (usually 50 or more tones).
    - Press **Menu** to confirm your selection, and exit to save the setting.
4. **Activate and Test**:
    - Once activated, your radio will only open the squelch for transmissions carrying the same CTCSS tone.
    - Test it with another Baofeng on the same frequency and tone to ensure that you only hear each other.

For DCS, follow similar steps, but choose **DCS** (Menu #12) in place of CTCSS. With more options than CTCSS, DCS is typically preferred in very crowded channels or environments with electronic interference.

Example Scenario Using CTCSS/DCS

Imagine you're on a camping trip with family members, and you want to communicate without interference from other campers using Baofengs nearby. By selecting a specific CTCSS tone (e.g., 67 Hz), your group's radios will only open when this tone is sent, filtering out everyone else on the same frequency. This ensures that conversations are clear and minimizes interruptions from overlapping transmissions.

Repeater Access and Offset Frequencies

Repeaters are essential tools for extending the range of your Baofeng radio, especially in areas where direct, line-of-sight communication is limited by obstacles or distance. Repeaters work by receiving your transmission on one frequency and immediately retransmitting it on another, usually from a high vantage point like a hill or a tall building. Understanding how to access repeaters using offset frequencies is key to maximizing your range and ensuring reliable communication over long distances.

How Repeaters Work with Offset Frequencies

Repeaters use two frequencies: one to receive (input) and one to transmit (output). The difference between these two frequencies is called the **offset**. When you transmit, your radio automatically shifts to the repeater's input frequency. The repeater then rebroadcasts your signal on its output frequency.

Offset Example:

- Let's say a repeater's output frequency is **146.100 MHz**, and it has an offset of **+600 kHz**.
- Your radio will automatically shift to **146.700 MHz** (146.100 + 0.600) when transmitting, allowing you to access the repeater.

Setting Up a Repeater on Your Baofeng

1. **Determine the Repeater Frequency and Offset**:
    - You can find repeater information online, or from local amateur radio organizations.
2. **Set the Frequency**:
    - In **Frequency Mode**, enter the repeater's output frequency (the frequency you'll listen to).
3. **Set the Offset Direction and Frequency**:
    - Access the **Menu** and navigate to **Offset Frequency** (Menu #26).

- Enter the offset in kHz (e.g., **600** for a 600 kHz offset).
- Then, set the offset direction to + or - (Menu #25) depending on the repeater's configuration.
4. **Save to a Channel**:
    - Save this setup as a channel so that you can access the repeater quickly whenever you're in range.

Example: Using a Repeater in a Rural Area

Let's say you're hiking in a remote area, trying to communicate with someone across a valley or mountain. By programming your Baofeng to access a nearby repeater, you can transmit further than your radio's typical line-of-sight range. This capability becomes a lifesaver when traditional radio communication falls short due to natural barriers.

Dual-Watch and Dual Receive Functions

Dual-watch and dual receive functions are powerful tools for maintaining situational awareness. These functions allow you to monitor two frequencies simultaneously, so you never miss important updates on a secondary channel.

Dual-Watch vs. Dual Receive

- **Dual-Watch**: Dual-watch (or dual standby) monitors two channels, switching automatically to the active one when it detects a transmission. However, it can only receive on one channel at a time.
- **Dual Receive**: True dual receive allows you to listen to two channels simultaneously, though Baofeng radios generally offer only dual-watch, which is still highly useful for most users.

Setting Up Dual-Watch on Your Baofeng

1. **Enter Frequency Mode**:
    - Use **VFO/MR** to enter Frequency Mode if you're monitoring frequencies rather than channels.

2. **Set Primary and Secondary Frequencies**:
    - Enter the primary frequency on the **A** display (upper).
    - Set the secondary frequency on the **B** display (lower).
3. **Activate Dual-Watch**:
    - Press **Menu** and go to the **TDR** (Dual-Watch, Menu #7) option.
    - Set it to **ON** to enable dual-watch functionality.

With dual-watch enabled, your Baofeng will scan between both frequencies, automatically tuning into the active one.

Example Scenario: Dual-Watch in Action

Imagine you're coordinating a large event. You set one frequency to the team's channel and another to an emergency frequency in case of unexpected issues. By enabling dual-watch, you stay connected to your team while also keeping an ear on critical updates. This function ensures you're well-informed, allowing you to respond swiftly to any situation without constantly switching channels.

CTCSS/DCS, repeaters, and dual-watch functionalities are the building blocks of advanced radio communication. These features allow you to filter out noise, extend your range significantly, and stay connected across multiple channels for increased situational awareness. Whether you're managing a hiking expedition, navigating a city event, or preparing for emergency scenarios, these advanced settings make your Baofeng radio a powerful communication tool.

CHAPTER SIX

Communication Protocols and Best Practices

Congratulations on reaching one of the most crucial chapters of this handbook! By now, you're equipped with the technical skills to operate a Baofeng radio effectively. But there's an important part of radio communication that goes beyond equipment: **communication protocols and etiquette**. Knowing how to communicate professionally and efficiently can make a huge difference, especially when clarity and speed matter most. Let's cover essential radio etiquette, how to evaluate and improve your signal strength, and best practices for emergency communication. By mastering these techniques, you'll become a reliable communicator who others can count on.

Radio Etiquette and Protocols: Speaking Like a Pro

Good radio etiquette keeps conversations organized and clear, avoiding confusion and ensuring that critical information isn't lost. Just as with any other communication tool, a few best practices and protocols can help you sound professional and communicate effectively.

Essential Radio Etiquette Tips

1. **Keep it Brief and Clear**: Unlike a phone call, radio communication needs to be concise. Stick to the main points, avoid long pauses, and use clear language.
2. **Listen Before You Speak**: Before pressing the Push-to-Talk (PTT) button, take a moment to listen. This ensures you don't interrupt another ongoing conversation on the same channel.
3. **Use Call Signs or Identifiers**: Identifiers or call signs help everyone know who's speaking and to whom they're speaking. Use these consistently so everyone in your group can follow along easily. If you're coordinating a group, assigning each person a unique call sign can streamline communication.
4. **Use "Over" and "Out" Properly**:
    - **Over**: Signals that you've finished speaking and are waiting for a response.

- **Out**: Indicates the conversation has ended, and no response is expected.

For example: "This is Alpha-One, moving to checkpoint three. Over." Then, wait for a response. Once the exchange is finished, you'd say, "This is Alpha-One, Out."

5. **Avoid Unnecessary Jargon**: While it's tempting to use military-style language, avoid excessive jargon unless everyone is familiar with it. Use clear, simple language, and prioritize clarity.

Call Sign Example for a Hiking Group

Imagine you're leading a small hiking group, and each member has a radio. You might assign call signs based on people's initials or an agreed-upon name. For instance:

- **Leader**: "Alpha-One"
- **Second in Command**: "Bravo-One"
- **Other Members**: "Charlie-One," "Delta-One," etc.

Using call signs like these helps each person understand who is talking to whom, even in stressful situations. Here's an example of a radio exchange with call signs:

- **Alpha-One**: "Bravo-One, what's your status? Over."
- **Bravo-One**: "Alpha-One, all clear at checkpoint. Over."

Emergency Signaling

When urgency arises, understanding emergency signaling and protocols can be lifesaving. For emergencies, you'll want to use a standard distress call, such as:

- **Mayday**: Used for immediate, life-threatening emergencies (e.g., "Mayday, Mayday, this is Alpha-One, we need assistance.")
- **SOS Signal**: Use Morse code for "SOS" (... --- ...) in extreme emergencies, or state clearly that help is needed, such as "SOS, need emergency assistance at [location]."

Understanding Signal Reports and Signal Strength

Signal reports are used to assess the clarity and strength of transmissions, allowing users to identify and troubleshoot issues with reception. Let's break down the basics of signal strength and quality so that you can diagnose potential problems and improve clarity.

Signal Strength and Readability Scale

Signal reports generally follow a standard system called the **RS (Readability-Signal) Scale**:

1. **Readability (R)**:
    - **1** = Unreadable
    - **2** = Barely readable
    - **3** = Readable with difficulty
    - **4** = Readable with minimal difficulty
    - **5** = Perfectly readable
2. **Signal Strength (S)**:
    - **1-2** = Very weak signal
    - **3-4** = Weak signal
    - **5-6** = Fair signal
    - **7-8** = Good signal
    - **9** = Strong signal

When giving a signal report, you combine the two ratings. For example, if you hear someone clearly but they're faint, you might respond with, "You're a 5 by 3," meaning readability is perfect but signal strength is weak. A "5 by 9" is the ideal rating, indicating a strong and clear signal.

Example: Testing Signal Quality

If you're unsure about your signal clarity, it's helpful to ask for a signal report. Here's a sample exchange for testing signal quality:

- **Alpha-One**: "Bravo-One, this is Alpha-One. How do you read me? Over."

- **Bravo-One**: "Alpha-One, I read you 5 by 7, clear but slightly faint. Over."

With this feedback, you might adjust your position or check your antenna to see if you can improve signal strength.

Emergency Communication

In an emergency, efficient radio communication is paramount. Baofeng radios are invaluable in disaster scenarios, outdoor survival situations, or when cell networks fail. Let's walk through best practices and techniques for using your Baofeng in emergency situations.

Setting Up Emergency Frequencies

It's wise to program emergency frequencies into your radio in advance. These might include local repeater frequencies or specific emergency channels, depending on your location and the type of emergency.

- **Common Emergency Frequencies**:
    - **U.S. National Weather Service**: NOAA Weather Radio (162.400-162.550 MHz) provides emergency weather alerts.
    - **Local Repeater Frequencies**: Find the nearest repeaters online or through local amateur radio clubs. Program these frequencies for expanded range in emergencies.

Broadcasting an SOS Message

In an emergency, an **SOS message** or a **Mayday call** can alert others to your need for immediate assistance. Here's how to structure an emergency broadcast for clarity and urgency:

1. **Identify Yourself**:
    - "Mayday, Mayday, this is Alpha-One, located at [specific location]."
2. **State the Nature of Emergency**:
    - "I am requesting immediate assistance due to [injury, lost person, etc.]."
3. **Provide Location and Details**:

- Include specific details, like GPS coordinates or landmarks.
4. **Repeat Key Information**:
    - Repeating your call sign and nature of the emergency ensures your message is clear, even if parts of the transmission are missed.

Example of an Emergency Broadcast

Imagine you're lost in a dense forest and unable to find your way back:

- **Mayday Message**: "Mayday, Mayday, this is Alpha-One, I am lost and requesting immediate assistance. My last known location is near the river at GPS coordinates 37.7749° N, 122.4194° W. Anyone receiving, please respond."

This structured message covers all essential information, helping responders locate you as efficiently as possible.

Using Morse Code for SOS

In extreme cases where voice communication is impossible, you can use **Morse code** to signal an SOS message, which universally represents distress. To do this:

- **SOS** in Morse Code: ... --- ... (three short beeps, three long beeps, three short beeps).
- Repeat as necessary until help arrives.

Tips for Optimizing Communication Clarity and Reliability

Even in non-emergency scenarios, reliable and clear communication can prevent minor issues from escalating. Here are some quick pointers to improve transmission quality:

1. **Choose Open Areas**: Try to use your radio in open areas without large obstructions to reduce interference.
2. **Maintain Line of Sight**: If possible, position yourself where there's a clear line of sight between you and the receiving station for optimal range.

3. **Reduce Background Noise**: In noisy environments, try to speak directly into the microphone to minimize background noise, or consider using a noise-cancelling microphone accessory.
4. **Check Battery Levels**: Ensure your battery is fully charged before any long communication session, especially if you're heading into a situation where charging won't be possible.
5. **Position the Antenna Correctly**: Hold the antenna vertically, not horizontally, to improve signal quality. Using a quality aftermarket antenna can also significantly enhance your range and clarity.

Mastering radio etiquette, understanding signal strength, and knowing how to handle emergency communication are vital skills that take your Baofeng radio experience to the next level. Whether you're coordinating with a team, reporting your status, or calling for help, these practices ensure your messages are clear, professional, and actionable. Practice these protocols, and you'll quickly become a trusted communicator in any situation, capable of responding to both routine updates and critical emergencies with confidence.

CHAPTER SEVEN

Troubleshooting and Maintenance

Your Baofeng radio, like any piece of gear, will work best when it's well-maintained and you know how to solve common issues that can arise. Just like a car needs regular oil changes and tire rotations, your radio requires basic upkeep to stay in peak condition. This will guide you to troubleshooting common problems, performing regular maintenance, and keeping your radio's software and firmware up-to-date. With these tips, you'll be ready to resolve issues quickly and extend the lifespan of your Baofeng, so it's ready whenever you need it most.

Common Issues and Fixes

Let's start with a few common issues you might encounter and how to solve them. Knowing how to address these will help you stay calm and capable even in a tough situation.

Power Issues: Radio Won't Turn On

Symptom: Your radio won't turn on or frequently turns off by itself.

Potential Causes and Fixes:

1. **Battery Issues**: Check that your battery is charged and seated properly. If the battery is old or visibly damaged, it may need replacement.
    - **Solution**: Try charging the battery fully and ensure it's securely attached. If that doesn't work, try a new battery.
2. **Faulty Power Button**: In some cases, the power button may become unresponsive.
    - **Solution**: Test the button a few times. If it's not working consistently, consider having the button inspected or replacing the radio if it's under warranty.
3. **Battery Contacts**: Dust or corrosion on the battery contacts can prevent proper power transfer.
    - **Solution**: Clean the contacts with a dry cloth or a small brush (avoid any liquids). Be gentle to avoid damaging the contacts.

Audio Issues: No Sound or Poor Audio Quality

Symptom: Either you're not hearing audio, or the audio is unclear or distorted.

Potential Causes and Fixes:

1. **Volume Level**: Sometimes, it's as simple as checking the volume knob.
   - **Solution**: Adjust the volume knob up and down to see if audio comes through at higher levels.
2. **Speaker Blockage**: Dust, dirt, or debris can block the speaker.
   - **Solution**: Use a soft brush to clean the speaker grill gently. Avoid pressing too hard or using liquid cleaners.
3. **Earphones or External Mic Issues**: If you're using an external microphone or earphones, the connection could be loose or faulty.
   - **Solution**: Remove any external devices and test the audio through the radio's internal speaker. If audio quality improves, you may need to replace the accessory.

Reception Issues: Poor Signal or No Reception

Symptom: Your radio struggles to pick up signals, or transmissions are weak and distorted.

Potential Causes and Fixes:

1. **Antenna Placement and Condition**: A bent or damaged antenna can reduce reception quality.
   - **Solution**: Ensure the antenna is securely screwed in and fully extended. If it's damaged, consider replacing it with a high-gain antenna for improved reception.
2. **Environmental Interference**: Buildings, mountains, and other physical obstructions can disrupt signals.
   - **Solution**: Move to a higher or more open area if possible. Some users report improved range by stepping outside or moving away from metal structures.
3. **Frequency Interference**: Signals from other devices can interfere with reception.

- **Solution**: Try switching to a different frequency or channel. Avoid using frequencies that overlap with other devices, like Wi-Fi or other electronic equipment, if possible.

Programming Issues: Channels Aren't Saving Correctly

Symptom: After programming, your radio won't save channels or keeps reverting to default settings.

Potential Causes and Fixes:

1. **Incorrect Programming Steps**: If steps aren't followed correctly, the radio may not retain settings.
    - **Solution**: Carefully repeat the programming steps from scratch. Refer to Chapter 3 on manual programming or using CHIRP software.
2. **Software Glitch**: Sometimes the radio's software might experience a minor glitch.
    - **Solution**: Turn off the radio, remove the battery, wait a few moments, and reinsert it before turning the radio back on. This can often resolve minor software hiccups.

Regular Maintenance

Maintaining your Baofeng radio will extend its life and ensure it's always ready when you need it. Here's a guide on how to care for the battery, keep the unit clean, and protect the antenna.

Battery Care

The battery is one of the most critical parts of your radio. Proper care can help you avoid premature battery failure and maintain longer talk times.

1. **Avoid Overcharging**: Overcharging your battery can reduce its lifespan. It's best to disconnect the charger once the battery reaches 100%.
2. **Regularly Rotate Batteries**: If you have multiple batteries, rotate their use so they wear out evenly.

3. **Battery Storage**: Store batteries in a cool, dry place, ideally at around 50% charge if you're not using them for a while. Avoid exposure to extreme heat or cold, as this can shorten battery life.
4. **Replace Old Batteries**: When a battery begins to drain quickly, it's time to replace it. Lithium-ion batteries typically last between 300-500 charge cycles.

Cleaning Your Radio

Regularly cleaning your radio keeps it in top shape and reduces the chances of dust or dirt affecting its functionality.

1. **Use a Soft Brush**: Gently brush away dust and debris from the buttons, speaker, and microphone grill.
2. **Avoid Liquids**: Keep liquids away from the radio, as water or cleaning products can damage internal components.
3. **Inspect for Damage**: Look for cracks or broken components. Repair or replace parts as needed to prevent small issues from escalating.

Antenna Care

The antenna is essential for good signal quality, so taking care of it is crucial.

1. **Keep It Straight**: Avoid bending or over-tightening the antenna. A straight antenna is best for optimal signal reception.
2. **Inspect for Cracks**: Check the antenna regularly for any cracks or other signs of damage. Replace it if you notice significant wear.
3. **Consider an Upgrade**: A high-gain antenna can boost your signal range, especially useful in outdoor or low-signal environments. Look for an antenna compatible with your specific Baofeng model.

Software and Firmware Updates

Firmware updates can fix software bugs, improve functionality, and sometimes add new features to your Baofeng radio. Staying updated is one of the best ways to ensure reliable performance.

Updating Firmware

1. **Check for Updates**: Visit Baofeng's official website or check with authorized retailers to see if a firmware update is available for your model.
2. **Use the Proper Cable**: Firmware updates often require a specific programming cable (usually a USB-to-radio cable compatible with Baofeng radios).
3. **Follow Instructions Carefully**: Updating firmware is a sensitive process. Follow the official instructions closely to avoid bricking (damaging) your radio.
4. **Backup Settings**: Before updating firmware, consider backing up your settings (e.g., channel lists) in CHIRP or another compatible software. This will make it easier to restore settings if something goes wrong during the update.

Example: Updating Your Baofeng UV-5R Firmware Using CHIRP

1. **Connect**: Plug in your programming cable to both the radio and your computer.
2. **Open CHIRP**: Launch CHIRP software, select your radio model, and download the current configuration.
3. **Backup Data**: Save a copy of your current settings.
4. **Install Update**: Upload the new firmware file to the radio through CHIRP, following prompts. Avoid interrupting the process until it completes.
5. **Restore Settings**: Once the update finishes, restore your previous settings from the backup.

Troubleshooting Flowchart

Below is a troubleshooting flowchart to help you quickly identify and solve common issues.

1. **Does the Radio Power On?**
    - **Yes** → Proceed to the next step.

- No → Check battery status, contacts, and power button.
2. **Is the Audio Clear?**
    - Yes → Proceed to the next step.
    - No → Adjust volume, clean speaker, and check audio settings.
3. **Is the Reception Strong?**
    - Yes → Radio is functioning well.
    - No → Move to a clearer area, check antenna, and consider environmental factors.

This flowchart can serve as a quick reference whenever you encounter problems, helping you systematically rule out potential issues.

Regular maintenance and a good understanding of troubleshooting are keys to reliable radio performance. By addressing power, audio, reception, and programming issues proactively, you ensure that your Baofeng is always ready when you need it most. Proper care of batteries, antennas, and software keeps your equipment at its best. With these strategies, your radio is likely to last longer and work better, making you a confident and prepared communicator.

CHAPTER EIGHT

Practical Applications and Real-World Scenarios

Your Baofeng radio is more than just a tool for communication; it's a lifeline in various real-world scenarios where reliable contact is crucial. Let's go into specific uses for Baofeng radios in outdoor adventures, prepping for emergencies, and staying connected in urban and group environments. Whether you're a solo hiker, a family in disaster-prep mode, or organizing an event, these radios are invaluable for maintaining situational awareness and ensuring you're in touch when it matters most.

Outdoor Adventures: Hiking, Camping, and Wilderness Scenarios

For outdoor enthusiasts, staying connected in the backcountry is essential. With limited cell reception in remote areas, a Baofeng radio can become a key part of your adventure kit, helping you maintain contact with your group, alert others in case of emergencies, and even receive critical weather updates. Here's how Baofeng radios excel in outdoor settings.

1. Staying Connected with Your Group

Imagine you're hiking in a group but are spread out over rugged terrain. Using your Baofeng radio, you can quickly check in on each member and relay essential information, such as upcoming trail markers or navigational changes.

- **Example**: Say you're hiking up a mountain trail. The person at the front notices a steep patch ahead. They can use the Baofeng radio to alert those further behind about the challenging terrain and suggest taking it slow. This communication keeps everyone informed and safe.
- **Suggested Channel Setup**: Pre-program a channel specifically for your group before starting your hike. For instance, if there's a popular emergency frequency in the area, program it as a secondary channel, allowing you to quickly switch over if assistance is needed.

2. Emergency Signaling in the Wilderness

If an emergency arises while hiking or camping, such as someone getting injured or lost, having a radio enables you to signal for help effectively, even without cell service.

- **SOS Feature**: Program one of your radio channels with the national or local emergency frequency if available. Practice using this channel to ensure you know how to call for help if needed. In the U.S., try tuning into 146.520 MHz, the primary calling frequency for emergencies on VHF.
- **Repeater Access**: Research local repeaters near your campsite or trailhead. By tuning into these repeaters, you can extend your radio's range and potentially communicate with people far beyond your immediate area, increasing your chances of receiving help.

3. Receiving Weather Updates

Weather can change rapidly in the wilderness, and being caught off guard can be dangerous. Baofeng radios can pick up NOAA weather alerts, providing you with real-time information on incoming storms or hazardous conditions.

- **Programming Weather Channels**: Many Baofeng models allow you to program NOAA weather channels, which broadcast weather information 24/7. Having a weather channel ready to go can alert you to incoming storms, giving you enough time to find shelter.

Prepping and Emergency Situations

When preparing for emergencies, whether natural disasters, power outages, or unexpected situations, having a dependable communication tool can be a game-changer. Here, Baofeng radios offer reliability and flexibility for any prepper or individual looking to stay informed during a crisis.

1. Disaster Preparedness

In situations like hurricanes, earthquakes, or fires, traditional communication networks often go down or become overwhelmed. A Baofeng radio allows you to stay connected with family, friends, and neighbors, coordinate plans, and stay updated on the latest information.

- **Example**: After a hurricane, cell towers may be down, making it difficult to contact loved ones. By coordinating with your neighborhood to establish a specific channel for emergency communication, you can create a local support network, relay critical information, and request supplies or assistance.
- **Backup Battery Plan**: Always have backup batteries for your radio. In emergencies where power may be out for extended periods, a fully charged spare battery can make a big difference.

2. Monitoring Local Emergency Channels

During emergencies, first responders and other emergency personnel communicate on specific channels. By programming these into your radio (within legal limits), you can monitor ongoing situations and stay aware of changes.

- **Emergency Frequencies**: In the U.S., many public safety frequencies operate between 150-174 MHz on VHF. Check local frequencies for emergency broadcasts and program them in advance. This way, you can monitor local police, fire, and EMS channels, giving you insights into road closures, evacuation routes, and other important updates.
- **Channel Scanning**: Use the scan function to cycle through multiple programmed channels to monitor local activity. This is especially useful if you're coordinating with multiple groups or family members in different areas.

3. Establishing a Family Communication Plan

For families, creating a plan that includes designated channels and protocols for using Baofeng radios can be a lifesaver in high-stress scenarios.

- **Example**: Develop a family plan that includes where to meet, what channel to tune into, and specific times to check in. For instance, you could agree to check in on a designated frequency every hour if family members are separated.
- **Emergency Protocol**: Practice emergency drills with your family, so everyone knows how to use the radio, call for help, and communicate effectively in an emergency. Consider having laminated cards with simple instructions for younger family members or anyone unfamiliar with radio operation.

Urban and Group Use: Staying Connected in Urban Environments and Event Coordination

Baofeng radios aren't just for backwoods or disaster scenarios. They're also excellent for keeping in touch during urban outings or coordinating with larger groups at events. Whether you're at a festival, on a road trip, or coordinating a family reunion, Baofeng radios provide a reliable means of communication in crowded or dispersed situations.

1. Event Coordination

For group events like festivals, weddings, or parades, Baofeng radios allow organizers and participants to stay connected without relying on cellular networks, which can become unreliable in crowded areas.

- **Example**: Say you're helping coordinate a music festival. Assign each team (security, logistics, first aid) a specific channel. You can also have a primary coordination channel to streamline communications.
- **Channel Labels and Clear Protocols**: Establish clear channel assignments and protocols for using the radio. For example, Channel 1 might be for general coordination, while Channel 2 could be designated for security only. Labeling channels and enforcing radio etiquette keeps communications organized and efficient.

2. Family Communication in Crowded Areas

If you're at a crowded location like a theme park, museum, or sports event, keeping track of family members can be challenging, especially if cell signals are weak.

- **Example**: At an amusement park, each family member carries a radio and uses a designated channel for easy check-ins. This way, if someone becomes separated, they can quickly relay their location and meet-up point without relying on cellular service.
- **Battery Management**: Urban events often last all day, so it's essential to carry spare batteries or portable chargers to keep radios running. Consider setting intervals (e.g., every 30 minutes) to check in with family members.

3. Road Trips and Caravans

If you're traveling with multiple cars on a road trip, Baofeng radios are excellent for coordinating pit stops, ensuring everyone stays on the right route, and communicating in areas with limited cell service.

- **Example**: For a family road trip with multiple cars, each driver uses a Baofeng radio to stay in contact. When the lead driver sees a good rest area, they can relay the information to the other cars immediately.
- **Safety Check-Ins**: Establish a protocol for checking in at regular intervals, such as every 30 minutes or whenever you pass a specific landmark. This keeps everyone in sync, even if they temporarily lose sight of each other on the road.

You can see that Baofeng radios are far from just a "hobby" gadget. They're versatile tools that can keep you connected in countless real-world scenarios, from the backcountry to urban adventures and emergency situations. When you use them strategically, by planning channels, understanding emergency protocols, and coordinating with your group, they become invaluable assets. So, whether you're hiking a remote trail, preparing your family for a potential disaster, or organizing a large event, your Baofeng radio stands ready to ensure you're always within reach

CHAPTER NINE

Expanding Your Skills and Knowledge

So, you've spent time learning the basics of your Baofeng radio, understanding its features, and mastering the essential functions. But what's next? The world of radio communication is vast, and there's always more to explore and learn. Let's dive into how you can continue expanding your skills and knowledge, connect with the global radio community, obtain advanced licenses and equipment, and find ongoing learning resources to keep growing in this fascinating field.

Connecting with the Radio Community

One of the most rewarding aspects of mastering Baofeng radios and amateur radio in general is the opportunity to connect with a passionate community of like-minded individuals. Whether you're a beginner or an experienced operator; joining a community can be a huge boost to your learning and practical skills.

1. Why Connect with the Radio Community?

Ham radio operators often refer to themselves as part of a "community" rather than just a group of people with a shared interest. Connecting with others gives you access to a wealth of knowledge, real-world advice, and hands-on experiences that are invaluable for improving your own radio skills.

- **Learn from Others**: When you're part of a radio community, you can learn tips and tricks from seasoned operators. Whether it's optimizing your antenna setup, troubleshooting common issues, or understanding advanced radio techniques, the community is a goldmine for practical advice.
- **Event Participation**: Many amateur radio operators participate in organized events like Field Day, radio contests, and local radio nets. These events are excellent opportunities to practice your skills in real-world conditions, learn about new equipment, and meet other operators.

2. Finding Local Radio Communities

Getting involved in local ham radio groups is an excellent way to expand your knowledge. Look for local amateur radio clubs or groups that meet regularly to discuss topics of interest, share experiences, and offer mentoring for newer operators.

- **Example**: If you're based in a larger city, you can search online for "amateur radio clubs near me" or check websites like QRZ.com, where you can find local clubs and upcoming events. If you're in a rural area, don't hesitate to use online forums and networks to find virtual clubs or local meetups.
- **Radio Nets**: Many regions hold regular "nets" – scheduled communications on specific frequencies where operators check in and exchange information. These nets are often educational and can provide insight into how more experienced operators handle different scenarios.

3. Engaging Online Communities

If local options are sparse, there are plenty of online ham radio communities where you can interact with fellow enthusiasts from around the world.

- **QRZ.com**: This website is a hub for amateur radio operators, offering a large database of call signs, technical forums, and member pages. It's a great place to ask questions and engage with the community.
- **Reddit (r/amateurradio)**: Reddit hosts a popular amateur radio community where users discuss everything from beginner tips to advanced equipment. Here, you can find helpful threads, troubleshooting guides, and even real-time advice from experts.
- **Online Forums and Discord Servers**: Many radio enthusiasts use forums like eHam.net and QRZ Forums, or even specialized Discord servers, to chat, troubleshoot, and share ideas. These platforms are great for asking questions and exchanging knowledge.

Next-Level Licenses and Equipment

Once you're comfortable with your Baofeng radio, it may be time to think about upgrading your equipment and obtaining more advanced licenses. The transition from a basic operator to a seasoned ham radio operator opens up new possibilities in terms of frequencies, equipment, and skills.

1. Upgrading to Advanced Radios

While Baofeng radios are excellent for entry-level operators, more advanced radios offer expanded features, better build quality, and higher power output. Many ham radio operators eventually move to more sophisticated equipment to explore new bands, work with higher frequencies, and even operate in more challenging environments.

- **Example of Advanced Radios**: Consider upgrading to radios like the Yaesu FT-60R, Icom IC-7300, or Kenwood TS-2000. These radios provide access to additional frequency bands (like HF bands), enhanced sensitivity, and more reliable performance, especially in emergency or professional settings.
- **Why Upgrade?**: More advanced radios come with a wider range of features like full-spectrum monitoring, digital modes, satellite communication, and longer battery life. They are perfect if you plan to expand your skill set and take on more complex radio activities.

2. Obtaining a Higher-Level License

In many countries, the first step to becoming a licensed amateur radio operator is obtaining a Technician License. But that's just the beginning. The Technician License grants you access to VHF and UHF frequencies, but if you want to access more power, additional frequency bands, and advanced modes of operation, you'll need to upgrade to a General or Extra class license.

- **Technician License**: This is the entry-level license in the U.S. that allows operators to use the VHF/UHF bands and limited HF privileges. It's great for beginners, especially if you plan to stick with local communications.

- **General License**: The next level of licensing, the General License, opens up access to more HF bands and allows for long-distance communication. It's the go-to license for operators who want to expand their reach beyond local communications.
- **Extra License**: The Extra class license is the highest level of certification and grants access to all amateur radio bands. It's ideal for those who want to operate at the highest levels of the amateur radio spectrum, including advanced digital modes, high-power transmissions, and rare frequency bands.

3. Studying for the Exams

To obtain these higher-level licenses, you'll need to pass a written exam that covers topics such as operating procedures, radio theory, and regulations. Don't worry, there are plenty of resources available to help you prepare.

- **Books and Study Guides**: Popular study guides, such as *The ARRL Ham Radio License Manual* or *The Technician Class FCC License Exam Study Guide*, are excellent resources. These books break down the material in an easy-to-digest format.
- **Online Practice Tests**: Websites like *HamExam.org* and *QRZ.com* offer practice exams that simulate the real test. They help you assess your knowledge and focus on areas that need improvement.

Ongoing Learning Resources

Radio communication is an ever-evolving field. Whether you want to deepen your knowledge of radio theory, try out new operating modes, or learn about the latest technological advancements, there are numerous resources to help you stay up to date.

1. Books and Publications

Books are a cornerstone of radio learning. They cover everything from basic operations to advanced techniques. Some recommended reading includes:

- *The ARRL Handbook for Radio Communications*: A comprehensive guide that covers radio theory, design, and operation for all levels of experience.

- *The ARRL Operating Manual*: A focused look at how to operate your radio effectively, including tips for dealing with interference, managing signal reports, and using digital modes.
- *Emergency Radio Communications Handbook*: This book covers how to use radio for emergency preparedness and provides strategies for staying connected in disaster scenarios.

2. Online Courses and Tutorials

With the internet at your fingertips, online courses and tutorials offer an interactive and flexible way to learn more about radio communication.

- **ARRL Online Courses**: The American Radio Relay League (ARRL) offers online courses for every level of amateur radio operator. These range from basic introduction courses to more advanced certifications.
- **YouTube Channels**: Channels like "Ham Radio Crash Course" provide excellent video tutorials for beginners and intermediate operators. They cover everything from programming your radio to setting up antennas and troubleshooting common issues.
- **Webinars and Virtual Classes**: Many amateur radio organizations host live webinars and virtual workshops that can deepen your understanding of the technical and operational aspects of the hobby.

3. Forums and Online Communities

Engage with other operators by asking questions, sharing knowledge, and participating in discussions.

- **eHam.net**: A forum where operators can review equipment, ask questions, and share tips about operating and maintaining radios.
- **QRZ Forums**: Another popular forum for operators to discuss technical aspects, ask questions about licensing, and connect with others in the ham radio community.

- **Discord and Slack**: Many operators have moved to Discord and Slack groups for real-time chats. These platforms are great for discussing technical problems, learning from other operators, and making friends in the radio community.

4. Local Radio Clubs and Events

Getting involved in your local amateur radio club is one of the best ways to deepen your knowledge. Most clubs offer regular meetings, hands-on activities, and mentor programs where experienced operators can guide you through advanced topics.

- **Field Day**: Every year, amateur radio operators participate in Field Day, an event where they set up and operate radios in emergency-style conditions. It's a great way to test your skills and learn more about the practical aspects of radio operation.
- **Local Swap Meets and Conventions**: Attending hamfests (radio conventions) or local swap meets allows you to test new equipment, purchase accessories, and interact with others in the community.

Expanding your skills and knowledge in amateur radio is an ongoing journey. From engaging with the community, obtaining higher-level licenses, and upgrading your equipment, to utilizing various learning resources, there's always something new to explore. Continually building on your foundation, you'll gain the confidence to handle more complex operations and explore a wider range of frequencies, modes, and equipment.

Remember, radio communication is as much about learning as it is about connecting. Stay curious, keep practicing, and enjoy being part of a vibrant, global community of enthusiasts. Whether you're operating locally or reaching across the globe, the possibilities are endless.

CONCLUSION

As we reach the final chapter of this guide, it's time to reflect on what you've learned and how you can apply it to master your Baofeng radio. Whether you're a beginner just starting out or someone eager to dive deeper into the world of amateur radio, this journey is all about growth, gaining skills, exploring new possibilities, and ensuring you're ready for any communication challenge.

Summary of Key Takeaways

By now, you should have a solid understanding of how to use your Baofeng radio for both everyday communication and emergency scenarios. Let's quickly recap the essential skills and best practices we've covered in this book:

1. **Understanding Your Baofeng Radio**:
    - You've learned about the features that make Baofeng radios so popular—affordable, versatile, and user-friendly.
    - We explored different models like the UV-5R and UV-82, each offering unique features for various needs, from outdoor adventures to prepping for emergencies.
    - You've gained insight into key radio concepts such as frequencies, channels, bandwidths, and modulation types, giving you a stronger grasp of how radios work.
2. **Programming and Basic Operation**:
    - We discussed how to program your radio manually and with CHIRP software, giving you flexibility in customizing your settings for different situations.
    - You learned how to operate your radio, adjust volume, scan channels, and use important shortcuts to improve your efficiency in both casual and emergency communications.
3. **Frequency Management**:
    - You've become proficient in understanding the VHF and UHF frequency ranges, allowing you to select the best frequencies for your communication needs, whether for local or long-range transmissions.

- You now know how to scan multiple frequencies and organize them into channel lists for different activities or locations, streamlining your operation.

4. **Advanced Settings**:
    - You've mastered features like CTCSS/DCS for privacy, understanding repeaters and offset frequencies, and using dual-watch and dual-receive functions to stay aware of multiple conversations at once.
    - These advanced features will help you navigate complex radio environments with confidence.

5. **Communication Protocols**:
    - You're now equipped with the knowledge of radio etiquette, including the proper use of call signs, signal reports, and emergency communication protocols.
    - In critical situations, you understand how to use your Baofeng radio for emergency signals, ensuring clear and reliable communication when every second counts.

6. **Troubleshooting and Maintenance**:
    - You know how to identify and fix common issues with power, audio, reception, and programming, making you more self-reliant when things go wrong.
    - You've learned best practices for maintaining your radio, including caring for your battery, cleaning your equipment, and keeping your software and firmware up-to-date.

7. **Practical Applications**:
    - You've discovered how to use your Baofeng radio in a variety of real-world scenarios—whether you're hiking in the wilderness, preparing for an emergency, or coordinating with friends in an urban setting.

8. **Expanding Your Skills**:
    - You've learned how to connect with the broader radio community, where you can exchange knowledge and ideas, join events, and even pursue higher-level licenses and equipment.
    - There's always more to learn, whether it's upgrading your radio gear, deepening your understanding of radio theory, or practicing with more advanced communication techniques.

While understanding the theory and features of your Baofeng radio is crucial, it's the hands-on practice and exploration that will truly make you a confident operator. I encourage you to experiment with different settings, try out new frequencies, and communicate with other operators as often as possible. Remember, the best way to improve your radio skills is through consistent practice. Here are a few tips to help you continue growing:

- **Join Local and Online Radio Communities**: Seek out local ham radio clubs, online forums, or radio nets where you can connect with fellow operators. Not only will you learn from others, but you'll also stay motivated to keep improving.
- **Participate in Field Days and Contests**: These events are a great opportunity to challenge yourself, test your skills in real-world conditions, and learn how to manage more complex scenarios.
- **Don't Be Afraid to Experiment**: Whether it's using new features like repeaters, trying out advanced techniques, or fine-tuning your equipment, the more you experiment, the more you'll learn. It's all about finding what works best for you.

Your Baofeng radio is a tool that grows with you as you expand your skills. Don't rush the process, enjoy the journey and embrace the learning opportunities that come with each new discovery.

With great power comes great responsibility. As you become more proficient with your Baofeng radio, it's important to remember that you are operating in a shared spectrum of communication. This means respecting others, following legal guidelines, and using your radio responsibly.

- **Abide by Legal Regulations**: Be sure to stay informed about local regulations and licensing requirements. In the U.S., for example, operating a radio without the proper license can result in fines and penalties. Always check the rules and make sure you're operating within legal parameters.
- **Respect Other Operators**: Always be courteous and respectful when communicating on the radio. Follow proper protocols, avoid excessive chatter, and use clear, concise language.

- **Use Your Radio for Good**: Remember that your radio is not just a communication tool—it can be a lifeline in emergencies. Use it responsibly, especially in times of crisis. Whether you're offering help during a natural disaster or relaying important information, your actions on the radio can have a significant impact.

By practicing responsible communication, you not only ensure that you're operating within legal bounds, but you also contribute to a positive and supportive radio community.

Congratulations on completing the *Baofeng Radio Bible*! You now have a solid foundation to confidently operate your Baofeng radio, whether for everyday communication, outdoor adventures, emergency preparedness, or connecting with the global radio community.

The journey doesn't end here, there's always more to learn, practice, and explore. I hope this book has ignited your passion for radio communication and equipped you with the skills to make the most of your Baofeng radio. Stay curious, keep experimenting, and always be ready to share your knowledge with others.

Happy communicating, and remember, when it matters most, you'll have the power of reliable communication in your hands.

www.ingramcontent.com/pod-product-compliance
Lightning Source LLC
Chambersburg PA
CBHW060434220526
45465CB00008B/3130